STEM in the MILITARY

CYBER TECH in Warfare

By Tanner Billings

Cavendish Square

Published in 2023 by Cavendish Square Publishing, LLC
2544 Clinton Street, Buffalo, NY 14224

Website: cavendishsq.com

Disclaimer: Portions of this work were originally authored by Judy Silverstein Gray and Taylor Baldwin Kiland and published as *Cyber Technology: Using Computers to Fight Terrorism*. All new material this edition authored by Tanner Billings.

Library of Congress Cataloging-in-Publication Data

Names: Billings, Tanner, author.
Title: Cyber tech in warfare / Tanner Billings.
Description: Buffalo, New York : Cavendish Square Publishing, [2023]. |
Series: Stem in the military | Includes index.
Identifiers: LCCN 2022032011 | ISBN 9781502665584 (library binding) | ISBN 9781502665577 (paperback) | ISBN 9781502665591 (ebook)
Subjects: LCSH: Cyberspace operations (Military science)--Juvenile literature. | Cyberterrorism--Juvenile literature. | Terrorism--United States--Prevention--Juvenile literature.
Classification: LCC U167.5.C92 B55 2023 | DDC 355.4/1--dc23/eng/20220719
LC record available at https://lccn.loc.gov/2022032011

Editor: Kate Mikoley
Copyeditors: Abby Young and Michele Suchomel-Casey
Designer: Andrea Davison-Bartolotta

The photographs in this book are used by permission and through the courtesy of: Cover Gorodenkoff/Shutterstock.com; p. 5 DVIDS/U.S. Air Force photo by Airman 1st Class Frankie D. Moore; p. 6 DVIDS/U.S. Air Force photo by J.M. Eddins Jr.; p. 8 Wikimedia Commons/File:General Paul M. Nakasone (NSA).jpg; p. 10 PW Photograph/Shutterstock.com; p. 12 (main) BongkarnGraphic/Shutterstock.com; p. 12 (binary code) Heymo/Shutterstock.com; p. 14 akilasaki/Shutterstock.com; p. 15 Everett Collection Inc/Alamy Stock Photo; p. 17 William Barton/Shutterstock.com; p. 20 Rob Kints/Shutterstock.com; p. 22 DVIDS/U.S. Air Force photo by Paul Shirk; p. 24 Maksim Shmeljov/Shutterstock.com; p. 26 Wikimedia Commons/File:Gas Centrifuges - 1984(14655893576).jpg; p. 27 Africa Studio/Shutterstock.com; p. 28 Wikimedia Commons/File:Equifax HQ.jpg; p. 30 DVIDS/U.S. Cyber Command, Integrated Intelligence Center, Joint Operations Center by Josef Cole; p. 32 Wikimedia Commons/File:National Security Agency headquarters, Fort Meade, Maryland.jpg; p. 35 Wikimedia Commons/Edward-Snowden-FOPF-2014.jpg; p. 36 goodluz/Shutterstock.com; pp. 38, 39 DVIDS/Photo by Steven Stover.

CPSIA compliance information: Batch #CWCSQ23: For further information, contact Cavendish Square Publishing LLC at 1-877-980-4450.

Printed in the United States of America

CONTENTS

INTRODUCTION

Since the U.S. Army's establishment in 1775—and later, the official creation of the U.S. military in 1789—many have fought to protect the country. Soldiers have risked their lives on the front lines. Sailors have fought for their country at sea, and airmen and airwomen have protected people from the sky.

Today, these are still important jobs. However, there's much more involved in protecting the people of today's United States. For one thing, the military is now also tasked with defending the country's computer systems from attack. These days, nearly anything you need to know can be found online. Our entire society is connected by computers. Damage to information networks can affect a nation's economy and functionality. By examining previous network attacks, cyber-technology professionals can help prevent or mitigate attacks on computer systems. Known as cyber tech for short, cyber technology is a type of computer technology that specifically deals with the internet. In the military, this is a particularly important job because if sensitive information is compromised, the whole country could be at risk.

This is just one example of how cyber tech is an important part of modern warfare. Coding, which is how we communicate with computers to tell them what to do, is also an important

Cyber-tech experts often rely on their STEM skills to make sure the military's internet systems are as up to date and advanced as possible.

part of making sure our military computer systems work properly. STEM—science, technology, engineering, and math—experts are needed for many jobs in the military. Cyber technology is just one of many fields in the military that requires STEM experts to help protect the county.

The implications of **cybersecurity** extend to nearly every aspect of U.S. security. A good cybersecurity system can help strengthen the country's power grid, nuclear defense, and air traffic control system. Computer technology is evolving on a nearly daily basis, which is why those protecting our systems, known as cyberwarriors, must be extremely vigilant, staying one step ahead of enemies around the world who have the ability to attack the United States through the simple weapon of an internet connection.

Military members fighting cyberattacks must stay up to date on the latest technology in order to always stay on the cutting edge of cyber warfare.

CHAPTER 1

STAYING CYBER SECURE

Who, what, where, and why? These kinds of basic questions are often the first ones a cyberwarrior in the military must ask themselves when combating a **cyberattack**.

- Who carried out the attack or intrusion—and was it a nation state, a **hacktivist** group, or an individual?
- What information was accessed, and did it affect the entire system or just a portion? How was the data changed?
- Where was the source of the breach? Did it involve private computer networks, government networks, or classified data?
- Why did the attackers do it?

After the answers to these questions are known, the military can then figure out who has jurisdiction in the case and which agencies should be engaged in tackling the problem. Also important is figuring out who should be held accountable for the breach. How was the attack allowed to happen?

U.S. Cyber Command

The people responsible for answering all the questions about a cyberattack are often members of the United States Cyber Command (USCYBERCOM). USCYBERCOM was established in 2010 at the National Security Agency (NSA) headquarters at Fort George G. Meade, Maryland. Part of the U.S. military, the command's mission is to "direct, synchronize, and coordinate cyberspace planning and operations—to defend and advance national interests—in collaboration with domestic and international partners." The command's main areas of focus are defending the Department of Defense Information Network (DoDIN), supporting **combatant** commanders in carrying out their missions around the world, and strengthening the country's ability to withstand and respond to cyberattacks. Because the command's work crosses so many boundaries, the U.S. Army, Navy, Air Force, and Marine Corps all have components reporting to USCYBERCOM.

As of July 2022, General Paul M. Nakasone is commander of the U.S. Cyber Command and director of the National Security Agency. He began serving in these roles in May 2018.

The Three Basic Parts of Cybersecurity

Our world is wired and interconnected, and people and their devices are linked together. Yet sometimes, the interconnections are exactly where threats to security occur.

Computer network exploitation (CNE) is a form of spying because it uses a network to proactively extract and gather intelligence with significant financial, reputation, or political consequences.

A computer network attack (CNA) is similar but is designed to actually disrupt, deny, degrade, or destroy enemy networks, which could sabotage the ability to be productive or to access important information.

Computer network defense (CND) offers a strategy to deter cyberattacks and maintain confidence in the confidentiality, integrity, and availability of friendly networks such as phones, laptops, and software for users to access. Together, confidentiality, integrity, and availability are often referred to as the CIA triad because they are the three main components of information security and every information security program should be able to achieve at least one of the three elements.

What Is Hacking?

One of the most basic—but also most serious—types of cyberattacks is called hacking. You have probably heard about hackers and hacking. You might have even heard someone say they have been hacked. Hopefully you haven't been hacked yourself.

What does the term "hack" mean exactly? By definition, hackers are people who are proficient at using or programming a computer. They use their skills to gain unauthorized access to a computer network or file. The computer industry started using the slang term "hacker" in the 1960s as a nickname for someone who is particularly skilled at computer programming. It is likely derived from the term "hack," which means "to chop," or from the golf term "hacker," which means "an amateur player." Later, the term "hacker" came to mean someone in the computer industry who played harmless pranks or broke into computer networks for fun. Eventually, it meant someone who has more devious and illegal computer activity in mind.

"Black hat" hackers use their knowledge to breach or bypass an individual's or organization's internet security without permission. Sometimes, their efforts are coordinated with others. "White hat" hackers work on behalf of government organizations or companies to uncover vulnerabilities. These people are also

Not all hackers are bad. Companies and organizations often hire hackers to make sure their systems are secure.

called ethical hackers. Somewhere in between white hat and black hat hackers are "gray hat" hackers. They don't necessarily want to harm their victims, but they may hack into a system looking for vulnerabilities. They usually won't steal or expose information, but some will report the vulnerabilities to system owners, asking for money to fix the issues. Even if they don't leak any information or steal anything, this type of hacking is still illegal because they didn't get permission before attempting to hack the system.

Striving for Security

It's important to always keep your personal information secure. If hackers get ahold of a person's private information, such as banking information, the person could be in danger. At the very least, being hacked is a pain for any person to deal with. For the U.S. government and military, the risks could be even more severe. If enemies get ahold of certain information, the security of the entire country could be at risk. Luckily, the best of the best are on the job. Top cyber-tech experts work tirelessly to ensure U.S. government and military computer systems remain secure from hacking and other types of cyberattacks. In the event of a security breach, they have the knowledge and resources to act quickly.

Join the Debate

1. In the event of a cyberattack, what do you think is the most important question a person combating the attack must answer?

2. Do you think USCYBERCOM is an important part of the U.S. military? Why or why not?

3. Do you think gray hat hacking is ever ethical?

Put simply, a code is a set of instructions for a computer. Binary code uses zeros and ones to represent the computer's states of "off" and "on."

CHAPTER 2

EARLY CODES AND HACKING

When it comes to jobs in cyber technology, knowledge of coding is often a **prerequisite**. Coding, sometimes called computer programming, is an important part of protecting our computer networks and systems. Secure coding can remove vulnerabilities and prevent future cyberattacks from happening.

Secure coding may seem like a new technology, but different types of codes have been used throughout most of history. That's because a code is simply a system for sending messages using symbols or signals. When it comes to war, creating ways to conceal messages and ensure secrecy has always been important. Native Americans, ancient Greeks, the Chinese, and the Japanese and Germans during World War II (1939–1945) are just some groups that created early codes or secret forms of communication. Computer coding is just another method people have come up with to keep their communication secure.

Using a Native Language as Code

During World War II, radios and telephones were the most frequently used machines to transmit messages about battle plans and troop movements. It was difficult to keep these

Morse Code

As technology became more sophisticated, so did the means of transmitting messages. Once electricity was discovered in the 19th century, a new form of communication, called Morse code, was widely adopted. Developed by Samuel F. B. Morse in the 1830s, it uses a series of electric "pulses" that can be sent through a wire. The pulses are various combinations of short and long signals. The arrangements of these pulses—along with the silences in between them—translate into letters of the alphabet and numbers. This system was used extensively for transmitting messages on a radio before it was possible to communicate by voice using radios or telephones.

In 1851, a second type of Morse code was created in Europe. Called the International Morse code or the Continental Morse code, it simplified the original code and allowed for other letters with special marks.

International Morse Code

A	• —	N	— •	1	• — — — —
B	— • • •	O	— — —	2	• • — — —
C	— • — •	P	• — — •	3	• • • — —
D	— • •	Q	— — • —	4	• • • • —
E	•	R	• — •	5	• • • • •
F	• • — •	S	• • •	6	— • • • •
G	— — •	T	—	7	— — • • •
H	• • • •	U	• • —	8	— — — • •
I	• •	V	• • • —	9	— — — — •
J	• — — —	W	• — —	0	— — — — —
K	— • —	X	— • • —		
L	• — • •	Y	— • — —		
M	— —	Z	— — • •		

Morse code allowed people to quickly send and receive messages over long distances before they could do so by voice.

plans secret because the enemy could easily eavesdrop on the conversations. The U.S. military needed a code that could not be broken, so they chose Navajo, a Native American language few outside of the Navajo Nation understood. In 1942, the military recruited 400 Navajo to serve in the Marine Corps as "code talkers."

Private First Class Preston Toledo (*left*) and his cousin Private First Class Frank Toledo (*right*) use their native Navajo language to relay orders over a field radio for the U.S. Marines in 1943.

They developed and memorized an encrypted code based on the Navajo language. A short message could be encoded, transmitted, and decoded in 20 seconds. It was first used in the Battle of Iwo Jima in the Pacific, where hundreds of messages were sent and received—with no errors. The code was never deciphered by the Japanese, and the Navajo code talkers are credited with significantly assisting the Allied war effort. President George W. Bush presented the code talkers with the Congressional Gold Medal in 2001.

The Enigma

One of the most notoriously difficult code machines for the Allied coalition to break was called Enigma. Named for the word that means "a person or thing that is mysterious, puzzling, or difficult to understand," it was invented by a German engineer named Arthur Scherbius at the end of World War I (1914–1918). Nazi Germany adopted it for use during World War II. The code used to decipher a message changed every day. German operators would be given a plaintext message to encrypt. The operator would depress each letter key, which was scrambled to produce a random substitute, based upon the unique wiring of the machine. Each subsequent letter that was pressed—or entered—into the machine would advance the rotor slightly and activate a different electrical pathway. Each one of these pathways would produce a different letter substitution. Each letter represented an entirely new code. Although it looked like a series of randomly assembled letters, it was an easily transmitted message.

The German operator who decoded messages needed an Enigma machine with the same synchronized settings so that the process could be reversed. That helped to accurately decipher the plaintext message.

The key to maintaining the secrecy of the Enigma machines was a list of daily key settings and instruction documents, which were heavily guarded by the Germans. The code and the machines were—for a time—considered to be unbreakable.

Working at Britain's code-breaking center in an old mansion about 50 miles (80 kilometers) northwest of London, British computer scientist Alan Turing and his team of engineers worked feverishly to devise techniques to break the code. This included their own electromechanical device, and it identified the daily settings on the Enigma machines. They were able to crack the code in 1941 and intercept German messages. This information helped enable Allied forces to defeat the Nazis. Some historians have said that the work accomplished by Turing and his team shortened the war by two to four years.

An Enigma machine is shown here at the Alan Turing Institute in London, England.

Deep-Sea Hacking

During the Cold War, spying and covert intelligence gathering by both the Soviet Union and the United States was rampant. Nuclear submarines were a critical asset used during the Cold War to track other ship movements, maintain a nuclear presence in international waters, and intercept communications by the enemy. U.S. submarines did this by tapping into top-secret undersea telephone cables that the Soviet Union's military laid on the floor of the Barents Sea. In the book *Blind Man's Bluff: The Untold Story of American Submarine* ***Espionage***, authors Christopher Drew and Sherry Sontag revealed how the U.S. Navy "bugged" these cables with secret listening devices in the 1960s, allowing them to gather vitally important information on Soviet military operations. Special warfare divers aboard submarines made daring deep-water dives to attach the listening devices to the undersea cables on the ocean floor. U.S. naval intelligence officials were able to identify Soviet military plans. You could call it a very early form of hacking.

Spreading Viruses

Computer hacking became a greater threat when computers became connected to each other through the vast network we now call the internet. When it was first created, only government employees were using the system, then called ARPANET. Initially, it was named for the government agency where it was created: DARPA (the Defense Advanced Research Project Agency). With such a small number of people using ARPANET, most of whom knew each other, there was no real perceived need or demand for security. In March 1971, that all changed.

ARPANET users arrived at work one morning and found a mysterious message on their computer screens saying, "I'm the creeper. Catch me if you can!" It was an experiment, run by a computer engineer named Bob Thomas who wanted to test the theory that data can duplicate itself. The "Creeper" spread to 10 computers, causing havoc. It was called a "virus" because it spread like an illness, "infecting" the entire computer system. Although this was the first time an unidentified person placed malicious computer code onto the newly formed internet, it was certainly not the last.

In 1988, a graduate student at Cornell University named Robert Morris Jr. wrote an experimental, self-replicating computer program and inserted it into the internet. At the time, it was still a system used mostly by the government and universities. Morris made the program seem as if it originated at the Massachusetts Institute of Technology (MIT) to hide its real origins. Morris claimed he created the virus to see how big the internet was. The program spread like a wildfire, replicating quickly and infecting many university, military, and medical research machines, slowing them down or crashing them. Called the Morris worm, it was estimated to have affected 2,000 computers within 15 hours. Morris was convicted of computer fraud and abuse.

One year after Morris's virus, in October 1989, an antinuclear group hacked into the computers of the U.S. Department of Energy and the National Aeronautics and Space Administration (NASA). When employees logged into their computers, they were greeted with a screen that read, "Worms Against Nuclear Killers ... Your System Has Been Officially WANKed." The group had decided to take their political protests online. This was one of the first instances of what would become known as "hacktivism."

Soon, militaries, governments, and corporations around the world started to wake up to the fact that these individuals and groups could cause real damage to equipment and productivity. Governments recognized that hackers were not just motivated by making money. Hacktivism gave these groups notoriety and a certain degree of power and control. That's when governments, the military, and organizations realized accessing sensitive and secret information was a digital form of espionage.

One of the most well-known hacktivist groups works under the name Anonymous. In public, members often wear this type of mask, called a Guy Fawkes mask.

Since the "Worms Against Nuclear Killers" hack, there have been countless other acts of hacktivism. People have differing opinions on hacktivist groups. Some liken them to Robin Hood and say that the crimes they may be carrying out are for the greater good of certain causes. Others call them "cyberterrorists" and say that there is no place for their crimes. It's also important to note that the causes hacktivist groups take up can differ greatly, and the outcomes of the hackings differ as well. Much like the different types of "hat" hackers, it's not always black and white, or good and bad. There are gray areas too.

Join the Debate

1. Why do you think the Navajo language worked so well as a code in World War II?
2. How do you think early codes relate to modern computer coding?
3. Do you think there is ever a time when hacktivism is OK? Think of examples why or why not.

Technical Sergeant David Mooers and Senior Airman Mario Lunato are seen here working on a server in the 557th Weather Wing enterprise at Offutt Air Force Base in Nebraska. The Weather Wing uses cyber technology to provide analyses and forecasts.

CHAPTER 3

ATTACKS ON NETWORKS

Today, it's not only battlefields where war is fought. Cyberspace has grown significantly over the last few decades. This generally positive advancement has had one major downside: a new opportunity for attack. Enemy nations no longer have to meet on a battlefield to fight face-to-face. Instead, war can now be waged from a comfortable seat behind a computer screen.

Malware, Phishing, and Botnets

Cyberattacks occur on both public and private computer networks around the world. Many involve malware, which is short for "malicious software." It is hostile or intrusive, and it is typically embedded in harmless or useful software that attaches itself to the user's computer. Hackers use a variety of methods to break into a computer or network, including "phishing." In a phishing expedition, a hacker will send an email that appears to come from a friend or colleague with a request to open an attachment or click on a link. Once the recipient opens or clicks, malware invades the computer and tracks the user's behavior on the computer, extracts private information, or shuts down the computer completely. It can even force the computer to perform tasks without the user's knowledge. This is how "botnets" are formed.

In war, computers are used to surveil battle areas, launch weapons, and communicate. These same systems are at risk if an enemy breaches the network.

Botnets are a collection of many computers infected with malware that follow the demands of a malicious third party. These infected computers are often used to send repetitive tasks or spam emails. Some botnets have included more than 500,000 computers, all controlled by one person. Botnets can be used to spy on people, spam computers, steal private or confidential information, or shut down computers. These armies of hijacked computers can attack even more computers and networks, creating a flood of requests to a particular network or website. Overwhelmed, the targeted network or website can crash or deny legitimate users access. This is called a distributed denial of service (DDoS) attack.

A DDoS attack occurred in 2007 against the tiny Baltic country of Estonia, formerly part of the Soviet Union. The attack shut down the websites of the nation's newspapers, banks, the president and parliament, and various government ministries. It was devastating because Estonia is particularly advanced in its use of computers and networks and highly dependent on the functionality of its information technology networks. The Russian government was the suspected culprit, since the two countries had been recently engaged in disputes. Full of technology-savvy specialists, the people and government of Estonia were able to recover, but the cyberattack was a wake-up call for governments and militaries around the world. Safeguarding a nation's technology networks requires military-level protection—the same type of protection a military provides its nation's geographic borders.

Attack by Digital Weapon

The United States and its allies around the world are not always the intended targets of cyberattacks. In fact, one of the most severe attacks on a computer network is rumored to have been initiated by the United States (and Israel). Some experts even say it was the first time a cyberattack was used as a "digital weapon." The U.S. government and its military have not admitted to conducting cyberattacks against the nation's enemies. However, there is enough evidence to suggest that the 2010 Stuxnet virus, which targeted Iran's top-secret nuclear facilities and, specifically, its nuclear **centrifuges**, was likely carried out by the United States and Israel.

These centrifuges are sophisticated cylindrical machines used to enrich **uranium**, the first step in the process to produce nuclear weapons. The virus targeted the computer systems that control and monitor the speed of the centrifuges. Iran had been in pursuit of a nuclear weapons program, and the United States and its allies were eager to stop the nation's progress. The Stuxnet virus effectively shut down the centrifuges. It was the first time malware caused significant physical damage to a facility.

These gas centrifuges from a plant in Piketon, Ohio, look similar to the ones in Iran targeted by the 2010 Stuxnet virus.

A Virus or a Worm?

While the terms "virus" and "worm" are frequently used interchangeably to describe a computer program that infects and disables a computer, there are some subtle differences. Both are maliciously created by people who want to disrupt business, military, and government operations. A virus attaches itself to an executable file, requiring a human to open and run it. It travels from one computer to another, leaving infections along the way.

A worm is considered a subclass of a virus, but the difference is it doesn't require human action to spread. It can travel from one computer to another without human assistance because it takes advantage of information transportation systems located on the computer system. It is particularly dangerous because it can replicate itself hundreds or thousands of times, filling up a large amount of computer memory.

Computer repairs can be required after a system is infected with a virus.

Government Data Breaches

In June 2015, the U.S. Office of Personnel Management (OPM), the government agency that hires government employees and manages their security clearances, announced that hackers had broken into its computers, compromising more than 21 million records. Stolen information included Social Security numbers, addresses, and fingerprints. It was one of the largest breaches of government information ever. Investigators suspect the Chinese government of perpetrating the attack.

Data breaches don't just happen to government systems. In 2017, Equifax, a large credit reporting agency, announced its systems were breached. As a result, personal data—such as phone numbers, home addresses, and Social Security numbers—of 148 million Americans was compromised.

In 2020, there was another huge data breach of the U.S. government. Hackers, who were suspected to be part of a Russian intelligence agency, broke into a variety of government networks, including email systems of the Treasury and Commerce Departments.

When these types of attacks happen, the government and military want to act fast to protect sensitive data. However, it can take months to detect a breach. In this day and age, data breaches seem nearly impossible to avoid. That's why there are so many professionals whose jobs are centered around cybersecurity, preventing attacks when possible, and mitigating effects when attacks occur.

Join the Debate

1. How can you avoid spreading computer viruses?
2. Do you think protecting a country's computer systems is as important as protecting its physical land?
3. Do you think data breaches are inevitable?

A member of USCYBERCOM is shown here working out of the Joint Operations Center at Fort George G. Meade in Maryland.

CHAPTER 4

FIGHTING CYBERATTACKS

The United States—like many other countries—runs on computer systems. The country's power grid, nuclear power plants, water systems, transportation systems, online commerce systems, stock market, and cell phone networks all depend on computer networks to run. This means they're all at risk of cyberattack. Just as it is almost impossible to track every terrorist around the world who might try to injure or kill Americans, it is almost impossible to track every potential hacker and enemy nation who wants to harm the country. It takes a team of highly skilled and trained cyberwarriors to protect the nation from cyberattacks.

All About the NSA

The U.S. government's primary agency for the collection of information about cyberattackers is the NSA. Established in 1952, the NSA specializes in "signals intelligence," or SIGINT. This is intelligence gathering by the interception of signals—either communications between two people or electronic signals. The NSA monitors, collects, and processes all kinds of data for counterterrorism purposes. Some of its tactics are "passive," meaning it just listens to electronic communications, but it also maintains a physical presence all over the world where its Special Collections Service can insert eavesdropping devices in remote places. Allegedly, the NSA also engages in more aggressive tactics, such as close-in surveillance, burglary, and wiretapping.

In addition, the NSA is also tasked with protecting the integrity of U.S. government communications and information technology systems. In other words, it prevents the hacking of U.S. networks. It is a highly secretive organization. In fact, the government did not even acknowledge its existence for many years, and employees jokingly referred to it as the "No Such Agency." It operates as part of the Department of Defense, and the director of the NSA is also the commander of USCYBERCOM.

The NSA's headquarters are located at Fort George G. Meade, Maryland, along with USCYBERCOM.

Searching for Cyberterrorists

When it comes to protecting the nation against cyberterrorism, both offensive and defensive capabilities are required. Defending the nation's networks is important, but experts are also on the hunt for the people looking to attack the systems. Searching for terrorists throughout cyberspace is conducted by collecting information about them. However, this is only effective if adversaries don't know they are being watched and listened to. Otherwise, they will change their methods of communicating.

The eavesdropping conducted under the NSA's international surveillance authority results in the collection of tens of millions of telephone records daily—perhaps more—to track the location of mobile devices around the world. This bulk collection includes records of major telecommunications carriers based in the United States. The information provides data about the travel habits of telephone users and allows the NSA to create "target development" data. This is information that can lead to the identity of unknown associates of those targets the NSA does know about. Such monitoring of the communications of militants suspected of terrorism also includes accessing data from internet and social media companies based in the United States, including Facebook, Twitter, and Google.

Generally speaking, the NSA collects two types of information: content and **metadata**. Content is defined as the conversations that take place on telephone calls, through email, and in online chats. It also includes online browsing histories and other search activities. Metadata describes the amassed collection of information that the NSA defines as "information about content but not the content itself." An email would generate metadata information about the identities of the sender and receiver and the physical location of the sender. Similarly, telephone metadata would include the phone numbers of both the caller and receiver, as well as the length of the call. Often, the data would include the types of devices used for the conversation.

The Infamous Edward Snowden

The U.S. government employs many cyberwarriors who are entrusted to keep the nation's secrets—including information about adversaries and the methods used to collect it. The temptation to share secret collections of information led to a major news story. Edward Snowden became famous—or infamous—in 2013. Was he a **whistleblower**, traitor, or hero?

Opinions vary, but the facts are these: A low-level contractor working as a systems administrator for the NSA, Snowden had access to top-secret programs, and he leaked information about them to the media. He claims he was concerned about the

Metadata information becomes part of a vast database of devices and their locations, which allows the NSA to determine the habits of a cell phone user, as well as the phone numbers used on a regular basis. This helps the government uncover terrorist networks and allows it to track their activities.

On a daily basis, cyberwarriors in the military and at the NSA work in a security operations center (SOC), a type of facility where all their organization's information systems—websites, databases, networks, servers, desktops, and data centers—are monitored and defended. Those who work in analytical roles might spend their days going over huge amounts of metadata, looking for patterns and trends to reveal the locations, behavior, and intents of known and potential terrorists.

The Ethics of Cybersecurity

Monitoring online activity to prevent cyberattacks is seemingly an effective way to protect the nation from them. However, many government's collection methods and how it invaded the average person's privacy.

In response, a White House spokesman defended the bulk collection of data as an important tool to protect the country from terrorist attacks. President Barack Obama said, "Our nation's defense depends in part on the fidelity of those entrusted with our nation's secrets. If any individual who objects to government policy can take it in their own hands to publicly disclose classified information, then we will not be able to keep our people safe, or conduct foreign policy."

Snowden fled the United States to live in **exile** in Russia after he leaked information in 2013. Some saw his actions as treason. Others thought they were justified.

people have spoken out against such monitoring. They've called it an invasion of privacy, when average citizens feel they are constantly being "watched" by the NSA. Some even think this type of surveillance is unethical. Others think this is a necessary part of protecting the country from potential cyberterrorists.

Join the Debate

1. What type of backup systems do you think there should be for the computer networks our country runs on?
2. Do you think Edward Snowden was at all justified in his actions? Why or why not?
3. Do you think the type of surveillance carried out by the NSA is ethical? Why or why not?

Hands-on computer experience, along with critical-thinking and problem-solving skills, are necessary for landing a job in cyber technology.

CHAPTER 5

A FUTURE IN CYBER TECH

Do you love learning about STEM subjects? Do you spend hours learning about different computer systems? Is coding your favorite hobby? If so, you may have a future in cyber tech!

As cyberattacks become more widespread and sophisticated, the government and private companies alike are looking to hire more and more people to work on cybersecurity. These people need to be highly skilled in information technology and coding. In addition, these organizations seek dedicated professionals with hands-on experience to work on their most secure systems and networks.

Jobs in the Military

In addition to hands-on computer experience, those being hired by the military and other government agencies, such as the NSA, demonstrate problem-solving skills, pattern recognition, space perception, and critical thinking. Recruiters are looking for candidates who not only understand how to program computer code, but also know how to independently, patiently, and persistently troubleshoot a computer problem. Like a good detective, cyber specialists identify and follow clues to find

Get Involved!

Becoming proficient in cyber technology takes more than good video-gaming skills. Learning to code is often the first step in becoming a cyberwarrior. Look for classes in your community where like-minded individuals learn these skills together. You can also see if your school has a robotics club, where you can learn more about computer programming with your friends. If your school doesn't have one, talk to a teacher about starting one.

Once you've learned about coding, you might want to take part in a hackathon, cyber camp, or cyber challenge. Also called a hack day, a hackfest, or a codefest, a hackathon is an event where computer programmers collaborate on software projects—often lasting between a day and a week. Sometimes they are held as educational forums, and sometimes they are held with the purpose of creating usable software. HackingEDU, held annually in San Mateo, California, is the world's largest educational hackathon. Major League Hacking is the official student hackathon league, providing support and mentorship to more than 70 collegiate hackathons every year.

Cyber camps can be a great way to learn about computer programming. These students are participating in a camp led by a member of the 780th Military Intelligence Brigade.

threats and perpetrators. They love to break open computers and electronic devices, see how they work, and then reassemble them into working devices. If this description sounds like you, then a career in cybersecurity might be in your future.

Those who want to be cybersecurity experts for the military must also demonstrate a loyalty to the nation and willingness to defend it against all enemies—foreign and domestic. Many in the military recognize the cyber battlefield will force countries to maintain vigilance against threats, while also developing a technological advantage. Cybersecurity demands staying one technical step ahead of adversaries and anticipating how and where they might attack next.

Training young cyberwarriors to anticipate threats, decrease vulnerabilities, and remain adaptable may be the best defense to avoid future attacks.

Educating the next generation of cybersecurity experts is a high priority for military leaders. The NSA offers educational programs for students to promote math, science, and language education at the elementary, middle, and high school levels. The NSA and the Department of Homeland Security (DHS) jointly sponsor the National Centers of Academic Excellence in Cyber Defense and

CYBER TECH TIMELINE

1830s: Samuel Morse develops a telegraph code for communications based on electric pulses.

1941: British engineer Alan Turing breaks the German Enigma machine's code.

1942: Navajo Indians are recruited by the U.S. military to develop a secret code based on their native language.

1952: The National Security Agency is established.

1960s: U.S. attaches "bugs" to Soviet undersea cables, secretly collecting information on its military operations.

1971: The Creeper virus infects 10 computers at the Defense Advanced Research Project Agency (DARPA).

1988: The Morris worm affects 2,000 computers within 15 hours.

1989: An antinuclear group infects Department of Energy and NASA computers in protest. It is one of the first instances of hacktivism.

2007: The country of Estonia's government networks are anonymously attacked, shutting down the websites of its president, parliament, newspapers, banks, and various government ministries.

2010: U.S. military establishes the U.S. Cyber Command (USCYBERCOM).

The Stuxnet virus shuts down Iranian nuclear centrifuges.

2015: Computers at the U.S. Office of Personnel Management (OPM) are hacked, compromising more than 21 million records.

2017: Equifax systems are breached, compromising personal data of 148 million Americans.

2018: General Paul M. Nakasone begins serving as commander of the U.S. Cyber Command and director of the National Security Agency.

2020: The U.S. government is the target of a data breach likely carried out by Russian intelligence agencies.

This timeline shows key dates in the history of cyber technology.

Cyber Operations programs. Many universities around the country offer these NSA- and DHS-**accredited** degree programs.

Looking Ahead

Cybersecurity experts think it is possible for an even more serious attack on the nation's computer systems than has already happened to occur. If the country isn't prepared for such an attack, it could be devastating. That's why the U.S. military works hard to ensure our "cyber weapons" are ready to strike back at any moment. This requires having an exceptional workforce to support the mission. Luckily the workforce, and the pool of potential candidates for it, is growing rapidly.

The military looks for cyber experts who are technically adept, agile, and creative, in addition to being good coders and good problem solvers. As technology evolves, the need for skilled employees will continue to grow, and the skills they will be required to have may change. As the world of cybersecurity changes, having a well-trained cyber force will help the country stay ahead of evolving risks. Will you be part of the next team of cyber experts tasked with protecting the country's most sensitive computer systems?

Join the Debate

1. Would you want a job in the military working on cybersecurity? Why or why not?
2. What do you think is the most important step to take to prepare for a job in cyber tech?
3. What skills, if any, do you have that would make you a good fit for a job working with computers in the government?

GLOSSARY

accredited: Given official approval by a certain group or organization for something.

centrifuge: A machine that separates substances by spinning them around very fast, causing the heavier substances to move to the sides or bottom of the machine.

combatant: A person or group fighting a war or battle.

cyberattack: A hostile act using a computer to disrupt, disable, or destroy another computer or network.

cybersecurity: Actions taken to provide safety and security for information and the network where it is stored.

espionage: The act of spying.

exile: A situation in which someone is forced to leave their country or home and live elsewhere.

hacktivist: A person who breaks into a computer for political reasons or social purposes.

metadata: Data that provides information about other data; the word was first used in 1983.

prerequisite: Something you must have or have done before you can do something else.

uranium: A radioactive element used to make nuclear weapons and energy.

whistleblower: A person who reveals a secret. Often used to describe an employee who brings a perceived wrongdoing by their employer to the public's attention.

FIND OUT MORE

Books

Farndon, John. *Stickmen's Guide to Technology*. Minneapolis, MN: Hungry Tomato, 2019.

Gordon, Sherri Mabry. *Internet Security and You*. New York, NY: The Rosen Publishing Group, 2020.

Mapua, Jeff. *Working with Tech in the Military*. New York, NY: The Rosen Publishing Group, 2021.

Miller, Michael. *Cyberspies: Inside the World of Hacking, Online Privacy, and Cyberterrorism*. Minneapolis, MN: Twenty-First Century Books, 2021.

Websites

Code.org
code.org
Code.org is an educational organization offering opportunities for K-12 students to learn computer science and coding skills.

What Is Cyberwarfare?
www.livescience.com/what-is-cyber-warfare
Read more about cyberwarfare on this website.

Organizations

Cybersecurity and Infrastructure Security Agency (CISA)
245 Murray Lane
Washington, DC 20528-0380
www.cisa.gov
CISA is a government agency that works to reduce threats to both cyber and physical infrastructure in the United States.

National Security Agency (NSA)
9800 Savage Road, Suite 6272
Fort George G. Meade, MD 20755-6000
www.nsa.gov
The NSA protects U.S. national security systems and works to advance cybersecurity.

Publisher's note to educators and parents: Our editors have carefully reviewed these websites to ensure that they are suitable for students. Many websites change frequently, however, and we cannot guarantee that a site's future contents will continue to meet our high standards of quality and educational value. Be advised that students should be closely supervised whenever they access the internet.

INDEX

ABOUT THE AUTHOR

Tanner Billings is a writer living in New York. He enjoys studying the history of the military and keeping up with modern military technologies. When he is not writing, he enjoys hiking and taking trips to national parks with his family and dogs.